A Wildlife Family Album

Library of Congress CIP Data: page 207.

A Wildlife Family Album

National Wildlife Federation

*T*oo astonished to be afraid, student kayakers find themselves face to tail with a humpback whale, one of six suddenly frolicking in Alaska's Prince William Sound. With a "whoosh!" the whale sounded, its blowhole "snapping shut like a hatch cover" as it slipped beneath the surface. Paddling furiously, the exuberant boatmen pursued the humpbacks until they disappeared in a sea of bubbles.

Sharing wildlife encounters like this one has been one of the joys of nearly 20 years of magazine publishing by the National Wildlife Federation and has led irresistibly to the album which you now hold in your hands. It is our salute of admiration and gratitude to the men and women who capture these once-in-a-lifetime moments for all of us.

Wildlife photographers take us to places most of us may never visit and show us things we would surely miss if we were there—because few of us would pay the price in time and discomfort. The blizzard that swept across northern Japan, encrusting the whooper swans pictured on the previous page, also numbed the hands of Teiji Saga as he adjusted the lens on his Hasselblad. The result is his exquisite evocation of what it means to be a swan in winter.

That is the photographers' gift to us. Each click of their cameras becomes our passport to the wild kingdom. To articulate the insights that come in these glimpses of true wildness, we turned to our writer-editor friend, David F. Robinson. In his sensitive introductions to the nine sections of the album, we celebrate the stirrings of new life, the risks survived in growing up, and the self-confidence of mature animals —their courtship fights and dances, their quest for food and shelter. Reluctantly we accept their death on nature's terms, as a necessary passage in the flow of energy wherein life must yield to life.

Photographers also make us see what we do to wild creatures when our worlds collide—when the oil spills, the flood waters rise, the bulldozer moves in. And they encourage us with pictures of caring people—researchers who live among sharks, gorillas, or cheetahs to find the facts we need if man and animal are to continue to share the earth. Often their findings help other scientists unravel the mysteries of human life as well.

Choosing for this album the most unforgettable photographs among thousands in our collection was arduous. We hope you will find in them the reward that drives a photographer from a warm bed to be at his icy post by first light: the delight of dwelling for a moment in the wildlife family circle where we are forever alien, yet inexplicably at home.

Alma Deane MacConomy
Editor

5

A Wildlife Family Album

A Time to be Born

*N*ew life! Nature's exuberant shout echoes into every cranny of our planet. Everywhere among the myriad creatures of earth, new life begins to stir. It throbs within the speckled egg and squirms just under the mother's tautened hide. It clings to the harsh places and runs riot where the living is easy. Nature's offspring rise and fall, their generations prance and fade, yet each generation takes from the lifefire within it a tiny spark to kindle the next. And for each new babe there is a moment when the spark ignites, the journey begins. There is a time to be born.

For the creature that will one day be born a mammal, the journey begins in the security of a womb. Nourishment is piped in, wastes are drained away, and neither fears nor dangers intrude upon the utter tranquility. Mother prairie dog (right) may face perils every day; surely she must remain vigilant every waking hour. But in the dark, warm world within her, all is calm and serene. Soon enough her pups will confront those same perils on their own.

Birth is sometimes roughly given. The giraffe is unceremoniously dumped on the ground from a height of five feet. Many insects know no parent; they simply hatch from an abandoned egg and get on with the business of living. And for those strange Australian birds we call the mallee fowl, there is no moss-lined cradle in the treetop, no bustling parent waiting with a beakful of food, no downy breast to snuggle into when the harsh winds blow or the somber clouds weep. The mallee chick must fight its way out of its egg and then struggle up to the light and air through a yard-deep mound of soil or compost. There its own parent is so busy tending the mound in which other eggs are still being incubated that the hatchling is not even noticed as it scurries off to fend for itself.

It's said that the egg is one of nature's most perfect designs. All that the next generation needs is stowed within—food, air, genetic blueprints, and that beginning spark. It is all the womb the embryo needs. For most of the egg-born—the birds, the amphibians, the reptiles, the insects—life's first problem is how to get out.

It may also be the last, if the weather is throwing a tantrum or a hungry predator waits nearby. But for those that survive, a time to be born is a journey begun, an album turned to its opening page.

(Overleaf) An eared grebe has apparently decided to settle down and add to the spring population of a quiet marshy lake. With some off-again-on-again help from her mate, the female may already have built as many as five nests this season, filled them with eggs, and abandoned each one before building the final nest. The many deserted eggs may help to keep predators away from the one that counts.

*T*hese greater flamingos (right) aren't just standing around on their knobby soda-straw legs admiring their eggs. They're changing the guard: one sits while the other feeds. To keep the eggs close to incubating temperature—93°F.—each pair has to get its act together or the eggs won't hatch. Yet so skittish are greater flamingos that a pair will abandon their precious annual egg if invaded by larger relatives, the marabou storks. In sharp contrast, the rufous hummingbird (below) dives fearlessly at foes two to three times its size, even a blackbird or chipmunk lurking too near its tiny eggs.

This ostrich parent (below) is not playing with its eggs or putting the nest in order. It is turning each egg so that the embryo won't stick to one side of the inside wall. Turning also ensures that all sides receive an equal amount of heat so that the chick will develop evenly. During incubation, the embryo also absorbs minerals from the shell, benefiting the tiny ostrich in two ways: by strengthening its bones and by weakening the rock-hard shell which the hatchling must finally crack open without help.

The gharial egg (left), unlike bird eggs, must not be turned over. Scientists studying this endangered member of the crocodile family always mark the top of the egg as they lift it from the sand and keep it upright. This precaution keeps the reptile inside from breaking away from the life-saving spot on the yolk which serves as its umbilical cord.

A baby bird looks helpless, but powerful muscles liberate it from the egg. Each hatchling thrusts its egg tooth —a hard white tip on the beak—against the inside wall, lunging and turning in a circle until the shell cracks in two. Hours after the hatching, precocial birds, like Canada goslings (below) leave the nest to forage with their parents. Green heron nestlings (left), like other altricial birds, must be cradled in a nest for weeks of care until they can fly. Their color against the grey stick nest keeps news of their arrival from predators and mother helps by chucking the telltale white shells.

A zebra can afford to give the birth of a colt her full attention. Yards away her stallion stands guard, ready to fight off any animal that might try to take advantage of the mare's helplessness. Minutes after she lies down, out comes a glistening birth sac with a foal inside (top). When the colt (above) pokes his way out of the sac, mother licks his whole head to clear the eyes, ears, nose, and mouth. She then nudges him up on spindly legs to nurse (right). Within days, the youngster learns to recognize his dam's smell and whinny, and her stripe pattern which is as distinctive as a fingerprint.

(Overleaf) In the fur seal family, life is more hectic because the bull is always stomping and bellowing to keep other bulls from luring his harem away. Each mother gives birth to just one ten-pound pup, but the bull may claim several dozen cows, so he ends up with a good-sized family.

The emergence of a butterfly is a solo performance—no parent waits nearby to encourage or protect it. Now in the fourth stage of its life cycle, the monarch makes its debut as an adult through the ruptured transparent case (above) in which it changed from a pupa into its velvety winged form (upper right). If a puff of wind should blow it to the ground in this limp, flightless state, a hungry mouse or even an ant could kill it. That's why a windless, sunny day is ideal for the unruffling of the soft wings (bottom right) which firm up quickly in the dry, warm air. In perfect flight readiness, this beauty can now flutter safely away (far right).

A Wildlife Family Album

Growing Up in the Family

*H*i, Mom! It's good to have a parent or two before facing that waiting world. And what a wonder-filled world it is. There are sticks and pebbles and siblings to nibble. There are doting parents to beg food from and snuggle against and hide behind. There are nestmates or litter-mates or youngsters in the neighborhood for companionship and a little horseplay when the growing gets rough. The early pages of life's album are aglow with scenes of apparent loving and caring and tenderness.

So it seems, and so it often is. Yet being young is serious business. There are life-and-death lessons to be learned—how to find food and, after the parental cornucopia shuts down, how to keep from becoming food when the family disbands and it's every creature for himself. Today the fuzzy brown Adélie penguin chick (right) need only tap the parental beak, and a fast-food service delivers a warm chowder of regurgitated fish; before long it will learn to catch its own fish or starve. Today it cozies up to a parent's breast and finds itself enwrapped in a brood pouch; soon enough it will face the awesome antarctic blasts with only its own fat and feathers for warmth. Today its padded parents take turns shielding it from the beaks of skuas, kelp gulls, and giant fulmars on the prowl; tomorrow the parents may look away for just the few seconds the marauders need to attack so they can keep their own youngsters fed.

So life's first pages hold both the pleasure and the peril of growing up, the lessons learned and the pratfalls endured and the bonds forged between parent and offspring. Perhaps we humans see something else: the grinding boredom of sitting in a tangle of twigs for a month or so with a nestmate as ugly as—well, as yourself. No wonder squabbles erupt in nest and den and burrow; no wonder the youngsters wander or fidget or pester their parents to wit's end.

Perhaps, though, a more important element escapes us. The sow bear cuffs her cubs not because they exasperate her but because they must behave to learn, and learn to survive. In the broodmates' bickerings, too, a pattern of survival is being worked out in favor of the fittest.

The family, then, is nature's teaching machine, and most of the brainier animals are born into one. But many creatures—most, in fact—grow up without one. A pity, we say; how fortunate we mammals are to have a family. But then, isn't it just like nature to devise another way?

(Overleaf) A baby elephant can do no wrong. All members of his large family protect him and put up with his comic capers. Mother, aunts, and even grand-mother may nurse him for years. When he charges at brothers, sisters, and cousins, they shove back, gently teaching him the rules. If he has a problem, it is what to do with his trunk. Finally he stops sucking on the end of it and learns to use it to drink.

$\mathcal{P}$ictures like these can be taken only when the mother bear has first been sedated while asleep in her winter den. Biologist Lynn Rogers (below) has done that hundreds of times so he can count, weigh, and tag the cubs. Rogers always tucks the entire black bear family back in quickly and closes the den. The sow resumes a deep sleep (not true hibernation) while the cubs nurse and nap in the warmth of her furry body. She doesn't eat or drink after entering the den, yet her milk fattens them from a birth weight of eight ounces to five pounds each by the time they leave the Minnesota den in April.

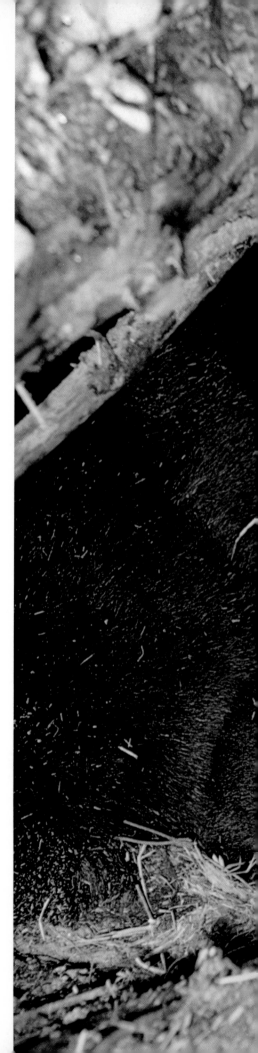

A mammal mother's gentle touch, scent, and voice create a bond with her offspring that is very important to their welfare. When the European dormouse (left) must go out to feed, her scent lingers in the nest, reassuring her blind, two-week-old litter and holding them there until she returns. An Alaskan moose (below) gives birth in a secluded area. This gives her twins a few days of safety in which to learn her voice and odor before they start following her. The bond between them becomes a lifeline when she must persuade the calves to swim across an icy lake to escape predators.

(Overleaf) Once a manatee calf finds the milk spigot hidden in a fold of skin under mother's flippers and learns to surface every few minutes to breathe, the living is easy. For two years the little marine mammal floats, nurses, and lets the old folks worry about sharks, killer whales, and motorboat propellers.

29

An armful of surprises has rewarded the scientists who are trying to domesticate the Arctic musk ox. The first surprise was the way the captured wild calves took to bottle feeding. Another was how quickly they learned new lessons such as what a fence is. Usually it took only one collision to convince them that free range was no more. When one ox showed others how to open a gate, the scientists knew they were working with intelligent animals. But instinct still operates, too. When barking dogs approach, even calves that never knew the wild form a defensive circle, just as their Ice Age ancestors did.

(Overleaf) Once a manatee calf finds the milk spigot hidden in a fold of skin under mother's flippers and learns to surface every few minutes to breathe, the living is easy. For two years the little aquatic mammal floats, nurses, and lets the old folks worry about sharks, killer whales, and motorboat propellers.

$\mathscr{A}$ fawn's joyful reunion with its mother (below) seldom lasts long, to reduce the chances of a predator tracking the doe to the odorless fawn's hiding place. Yet the doe's milk—twice as rich as a Jersey cow's—holds the fawn until she comes back four hours later. Baby hedgehogs (right) too dive in when mother returns from hunting, undeterred by the stiff spines on her back and sides. They know that her underside—like her face—is soft, hairy, and hospitable. A hungry little prairie dog (right, bottom) would probably agree that mother's milk is best, but substitutes are gratefully accepted.

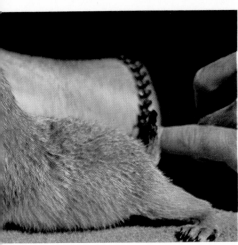

$\mathcal{O}$ne for you, and one for you . . . (left) No cedar waxwing parent is foolish enough to bring home one berry for four nestlings, nor to make four trips when one will do. Its crop serves as a fruit basket which can be filled with whole berries while out foraging. Then, back at the nest, the sight of each scarlet mouth brings the berries up one at a time.

The bond between a young robin and 5-year-old Rebecca Caccia (below) was also forged by food—earthworms to be exact. Her sculptor father takes in injured birds and the abandoned baby robin was one of the lucky ones.

(Overleaf) White pelicans handle food service a little differently than do the cedar waxwings. Parents swallow the fish they catch and digest them on the way home. There the downy chick must help itself. Reaching deep into the parental pouch, it gobbles more than a pound a day.

*T*his female American robin (below) has about had it. While she must help her mate yank up to ten feet of earthworms daily to feed their growing chicks, she must also stay at home more. Her job is to keep the naked birds warm until a coat of down followed by a coat of feathers helps establish their internal temperature controls. And now this nosy photographer is pushing the limits of her tolerance for uninvited guests. A more serene purple heron (left) performs the same guardian and temperature-regulating duties for her chick by acting as a heat shield in a sun-baked nest of reeds in Central Europe.

rched to spring, slash, and bite, a
bobcat mother (left) can make short
work of any fox or eagle foolish enough to
try to snatch her kitten. A slow-moving
opossum mother (below) confronted by a
young red-tailed hawk is virtually helpless,
but that doesn't stop her from making a
bold stand. She hisses and shows her 50
teeth—more than any other American land
mammal possesses. Her children are
mature enough to imitate this threat
behavior, but if the hawk closes in, they
are not likely to copy her final defense,
"playing dead." That instinct doesn't take
over until the young are on their own.

arching along to a slow four-four beat, a kid follows its nanny up a steep, snowy ridge in Washington's Olympic National Park. In the mountain goat family, kids have a year in which to learn from adults the whereabouts of grasses, sedges, and lichens—grazing plants scarce even in summer above the tree line. During this apprenticeship, kids imitate nanny's fierce rushes and other threat displays against mountain lions, wolves or lynxes, or aggressive billy goats who often pick on kids. The young goats practice on other kids by trying to force them down on their knees, or by playing "king of the mountain."

They look like hitchhikers, but each is actually at home. The koala riding piggyback (below) and the joey spilling out of mother kangaroo's pocket (right) are both convinced there's no place like mother. Each spent its first six months in the pouch, anchored to the milk supply and unaware of the dingo dogs and giant lizards that prey on their kind. At eight months the koala has outgrown the pouch, but the joey still squeezes in, turning a complete somersault each time it enters. Like most of Australia's 170 marsupial species, both face being weaned soon to plant food and independence.

Leaving the Nest

*O*ut of their cradles and nurseries the youngsters at last come tumbling, not quite ready to take on the world but more or less ready to try. It's a long way down for timorous little wood ducks (right) so new they've never been out of the nest box, but with a fluttering of tiny winglets and some gentle persuasion from mom, these two ducklings will yet make a splash. For weeks to come, they will paddle after her like skiffs abaft a schooner. But once they leap from the box, they leave the nest for good; there is no re-entry program. For a nest is not a home. It is merely a waystop on the journey from the egg to the outside world.

For many birds, the nest's last role is as a launching pad, a sky island for student pilots who can only learn to fly by flying—or trying. Thrashing the breeze like wind-up toys cranked too tight, they spill off their perches and flail frantically toward whatever twig looms ahead.

The journey from security to maturity can be long, tiresome, and at times a little scary. No wonder many young animals seem reluctant to get going. Who can blame the wolf pup for wanting its puppyhood to go on forever? The pack's grownups often snap and snarl at each other for the right to feed—but the pampered pups can usually scamper to the kill and fill up with impunity. After months of such spoiling, they'll need a few of those snaps and snarls when it's time to outgrow the good life and find a slot in the hierarchy of the pack.

Human parents hoarse from haranguing their own adolescents can all but hear the commands as haggard parents in many of nature's families seem to steer their offspring toward independence. Pity the young kangaroo that has lived for almost a year in its mother's pouch only to find the door to home sweet home slammed in its face.

Once-doting puffin parents evict not their young but themselves. Abandoned in a dark nest burrow, the puffling stays put until it has used up most of its baby fat. Heeding hunger's relentless prod, it finally leaves the burrow by night, walks to the sea, jumps in and teaches itself to swim, fish, and fly.

Pushed by their parents, pulled by their inborn curiosity, nature's newcomers lurch and scramble into the waiting world. As we chuckle at their pratfalls, their awkwardness, their first-time flops and try-again triumphs, do we not see in them our own brash and timid young selves?

(Overleaf) A fawn doesn't break the apron strings—it pulls at them gently. A week after birth, it may try to leave its hiding place to follow mother, but she restrains it, with a hoof if necessary. After she leaves, the young deer may wander off but returns when the doe calls with a sound like the mewing of a cat. After three weeks, they travel together. Even after weaning, it stays on to learn her survival secrets.

ere I come, ready or not—this one-day-old wood duck (right) answers mother's call to take its first swim, diving perhaps 50 feet to the pond or ground. Siblings who hesitate may find the family gone when they land, and their cries for help might attract instead a hungry raccoon. Another adventurer, a least bittern (bottom), can't fly but is already a good clinger and climber. Hanging in there against gravity in its cattail stand, it may not make it back to the nest. Young herons (left) freeze with pointed beaks when strangers approach. This still pose helps them blend with the background.

(Overleaf) After spending the first month of their lives in a cool underground nest, young burrowing owls find the sun's heat hard to take. To keep themselves cool enough for comfort, they open their beaks and pant like a dog. Since they can't fly, they stick close to the burrow, ready to scramble inside at mother's first sighting of a badger or a weasel.

Shaky test flights at about four weeks launched the flying career of this Eurasian little owl (above). Now fully fledged, the pigeon-sized little owl looks like the burrowing owl (pp. 54-55) in shape, coloring, habits, and flight. When hunting from a perch, the little owl zooms in on rats and rabbits in a brief, undulating, but silent, flight. On the ground, it scampers on long legs after bugs and beetles. Although this barn owl (left) can fly, it relies on strong body language for defense. Wings outstretched, it hunches, clicks its bill, hisses, and swings its head —an impressive show for a young owl.

(Overleaf) Adventure must wait awhile for ten-day-old egrets. To gain the strength and to grow the flight feathers they will need for their first takeoff over the Okefenokee Swamp waters 30 to 40 feet below, they devour all the snakes, frogs, and fish their parents bring them.

The living is never easy for downy, pearly-grey baby emperor penguins, but they get a lot of help the first five months. Parents take turns brooding the chicks for several weeks. Then they put them in a daycare center, called a crèche, and go off fishing together again. The parents continue to bring food to their own chicks until the young have moulted their down and grown waterproof feathers. Then the adults begin to ignore the young's open-beaked pleas for food. After several missed meals, the juveniles eventually get the message that the day of the free lunch is over and take the plunge into icy waters to fish for themselves.

(Overleaf) Young giraffes too are looked after by "aunties" who keep them with the herd while their mothers browse.
If danger looms, all adults show calves how to convert fear to a gallop. More than half of the calves die because they have not learned those two vital lessons.

*T*aking 10-to-12-foot vertical leaps to keep up with mother comes naturally to a young mountain goat in the Canadian Rockies (below). Cloven hooves with hard outer edges for digging in and tough cushioned inner pads for traction splay widely, giving the goat a sure hold even on slippery surfaces. A young raccoon (left) saves itself from a fall by relying on strong legs and sharp claws. Fear and hunger do wonders to improve the climbing skills of this roving bandito. But it's a big world out there and sometimes sheer wanderlust and curiosity make it impossible for young animals to stay home any longer.

A Wildlife Family Album

Coming of Age

Finally all of the pieces fit. The machinery of bone and muscle, the partnership of instinct and action drive the finished animal through its daily rounds with grace and competence and little wasted energy. Now, at last, the creature is equal to its world. It is what nature intended a whale to be, or an eagle; it is pelican or cuttlefish or caribou fully defined.

I am a bird, says the stunting acrobat. I am a porpoise, says the frolicking torpedo. I am a colt no longer, says the zebra. I pound the ground and rake the air and snort to all the plain: I am a stallion.

Now the humpback whale—that had to be shoved to the surface for its first gulp of air—comes vaulting out of the water (right) in an explosion of foam and blubber and ponderous high spirits. Now the peregrine falcon that entered the world as a bedraggled hatchling hurtles down upon its prey like the soul of a thunderbolt, surpassing 130 miles an hour, sometimes twice that. Now the giraffe attains its lofty overview, the golden pheasant wins its spurs, the firefly gains its blinking light.

For most animals, maturity means full size. They may live longer but they will get no larger. Not so the fishes; the longer they live, the bigger they get. And not so the chisel-like teeth of the bustling beaver; unless this workaholic woodchopper keeps felling trees all its life, its own mouth may be permanently propped open by its ever-growing incisors. But even the smallest grownups ask no quarter as they sally forth—the mosquito with its hypodermic needle, the jeweled stinkbug with its chemical arsenal, the stonefish that hides by looking like a stone on the ocean bed.

But among social animals, even those at the height of their powers sometime have to play second fiddle. One has to lead and the rest must follow. The leader is often the biggest and strongest, but there is more to leadership than age, size, or brawn. Chimp and wolf and musk ox soon learn that they must work their way up the ladder or split off and form a tribe of their own. The elephant herd enthrones not its biggest bully but a courageous sage, a wrinkled matriarch schooled by decades of experience in caring for her own.

Alone or in groups, mature animals of every kind move out to claim their place on the planet. Instinctively vigilant, yet as sure of themselves as they will ever be, they are in charge at last.

(Overleaf) Soaring over its Eden-like island, a young Galapagos hawk looks wild and free. Yet it is so unafraid it will take food from your hand. That is, if you don't mind offering it a live rat, dove, or lava lizard. At maturity, the buff-and-white plumage on its three-pound body all turns sooty brown. Then two males may share a mate and take turns bringing food to her and her chicks.

68

Exploding from the blue Pacific, a gray whale awakens all our childhood feelings about monsters of the deep. A once-in-a-lifetime click of the shutter preserves the moment it hangs there atop the wave it has churned up. Staring into the whale's cavernous maw, we marvel at the daring of photographer Howard Hall who swam in the green kelp for two hours, camera in hand, hoping for such a picture.

We can see the whale's baleen — a white food strainer hanging from its upper jaw — but we know this 21-foot yearling is not here to feed. These are warm California waters where it fasts as it migrates with other juveniles, trailing after the breeding adults and the cows that give birth here each winter. While the calves are still nursing, the whales all head back to cold arctic waters and their first meal in six months. There the baleens net small crustaceans by the ton to rebuild the whales' depleted stores of blubber.

What do these two bull giraffes know that the zebras grazing in their shadow don't know? Something— perhaps a skulking lioness—has caught the keen eyes of these 15-foot lookout towers. If their stilty legs break into a gallop, the zebras will whirl away with them without a backward glance. Next time, the lioness may succeed in stalking the giraffes until they lie down for a snooze. But even then, their way of dozing with head erect and each one facing a different direction may enable them to rise in time to stomp the lion to death.

(Overleaf) Yes, it's a crocodile. The fourth lower tooth that shows when the jaws are closed tell you it's not an alligator. A quick succession of bites by this Nile crocodile can crunch and crush the spine of a water buffalo—a task made easier by drowning the 1,500-pound buffalo first. Size, power, speed, and agility all justify the smirk on the croc's armor-plated face.

One step at a time, a bull walrus (right) and a bighorn sheep (left) solve the problem of getting around in the harsh terrains that they call home. The walrus' giant flippers propel it effortlessly in the sea, but if the male wants to soak up some sun or the female is ready to give birth, each needs the help of those two-foot ivory tusks to hoist its 2,000-pound body out of the water. In spite of its nickname, "tooth walker," the walrus does not use the tusks to move on land. Instead, its bent front flippers pull the massive hulk forward as it undulates along. But the chisel-like tusks double as a slashing weapon in fights for a mate or for the best place to nap in the sun.

Meanwhile, out on the Continental Divide, the sure-footed bighorns edge their way. On precarious paths inches wide, they climb and leap to reach the scarce grasses with the steadiness of bison grazing on the plains 10,000 feet below.

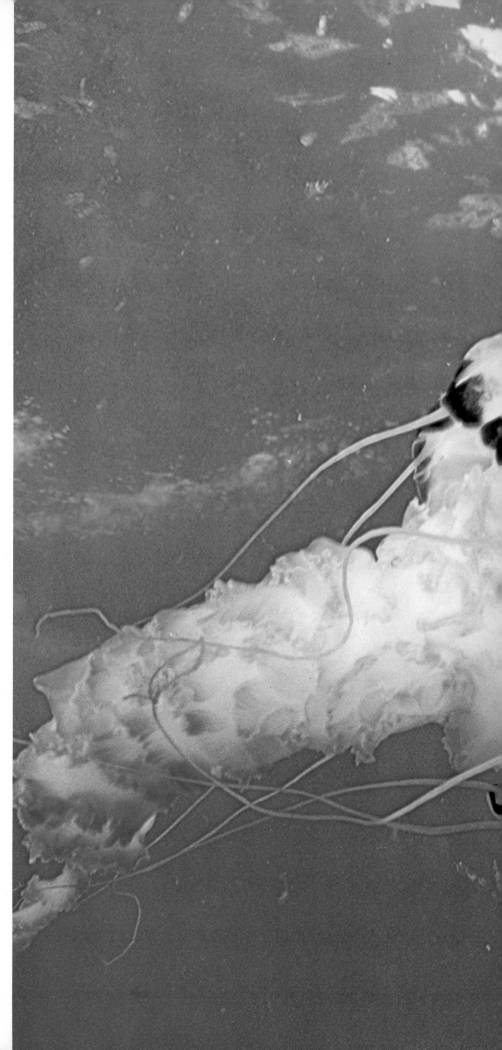

It looks like a parachute that has fallen into the sea, but it's really a pelagia jellyfish that lives among California's Channel Islands. A diver found it jetting along by enlarging and shrinking its canopy. Each contraction forces water out the lower end of its gelatinous body, giving it a forward boost.

Even in these warm waters, the diver has taken care to wear both wetsuit and gloves so he can enjoy the beauty of these exotic travelers without suffering from the sting of their tentacles. Fragile as spider webs, the lovely creatures are quite able to fend for themselves. At the slightest touch, a tentacle releases a hollow thread that injects poison into the toucher, whether prey or underwater sightseer.

Alive with the beating of a thousand wings, the Yukon River froths with restless mallards caught between the urge to stay and the urge to migrate. The secret of how they know when to go, and where, still lies hidden beneath the sheen of those magnificent green heads. We once thought the sun and the stars told them, until we realized they fly through cloudy days and storm-filled nights. Now we wonder if they are attuned to the earth's magnetic or gravitational fields. Scientists are still trying to unravel the mystery of how their internal compass works. We only know that millions of mallards pass over our heads twice a year, flying a mile high at 25 to 50 mph, each one living proof of the power, the mystery, and the beauty of flight.

The dazzling two-inch Guiana poison frog (right) flaunts its colors in the rain forest, reversing the defense pattern of most small creatures that hide in the daytime. The enlarged picture shows the psychedelic pattern that warns predators to shy away from the deadly secretion of its skin. Another South American native, the horned frog (below), lacks bold colors but confounds its enemies with bold action. This wide-mouthed ambusher of prey larger even than itself is so aggressive that predators shun it as if it were venomous.

Even so fragile a creature as the death head sphinx moth (left) takes a look-at-me approach by wearing a Halloween mask on its back. And if that doesn't stop a hungry bird, the sphinx has another unmothlike defense that may save it at the last minute: a loud squeak.

$\mathcal{W}$allflower, showoff, magician—the cuttlefish (below) is a world-class entertainer of divers in tropical waters. Left alone, it fades into the background by constantly changing color and pattern— from one hue to stripes or polka dots if needed. When hungry, the cuttlefish glows brilliantly as it overpowers its prey. But when attacked, it goes POOF, and makes a fast exit behind a blob of ink it has just expelled. Even the porcupine fish (right) would find the disappearing cuttlefish a hard act to follow. To scare off predators, it inflates its 22-inch-long body with water and air into a prickly ball 14 inches in diameter.

(Overleaf) An eight-inch coney is reasonably safe as long as it stays close to its matching coral hiding place in the Bahamas. But if it gets excited, it may blow its own cover. At the sight of predator or prey, the lower half of the coney's body turns white, the upper half, brown.

$\mathcal{K}$eeping warm in winter was no problem for macaques when they came to Honshu a million years ago. But the adaptable little primates refused to die out as volcanic upheavals and climatic shifts turned their warm island into the Japanese Alps. Instead, they developed thick, long-haired coats and became tree dwellers to escape the five-foot blankets of snow on the steep, forested slopes. The young learned to make snowballs (which they just carry around) and the adults discovered the luxury of soaking themselves in the hot sulphur springs. Today spas are set aside for their use in winter.

Grooming (below) is a great comfort, but permission to groom or to be groomed must be requested and may be refused. It is this kind of orderly cooperation that has enabled the snow monkeys to survive by huddling together in blizzards and warning each other of avalanches.

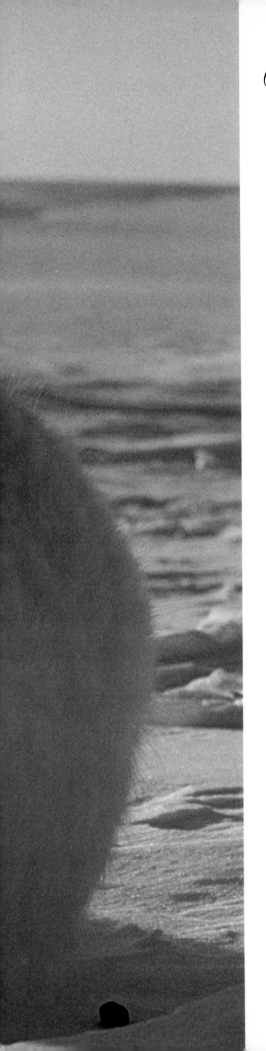

How can arctic animals look comfortable in some of the world's coldest weather? Because they have eaten enough to generate heat and little of that heat escapes through their bodies' insulation barriers. The hare (left) has a dense undercoat of fine hair, protected by long, coarse hairs that interlock when pressed. Its feathered neighbors, a trio of ptarmigan (bottom), dig roost holes in the snow.

Rocky Mountain elk (below) too must eat to keep warm, but they can also migrate —some of them down into the geyser-warmed valleys of Yellowstone National Park where a blizzard caught this pair.

(Overleaf) Africa's heat drives hippos into the cooling comfort of a mud wallow. The wallow grows as showers fill the depressions left by each mud-caked, two-ton body. When too rich with hippo manure, the wallow is abandoned. Plants take over, giving shade to smaller creatures.

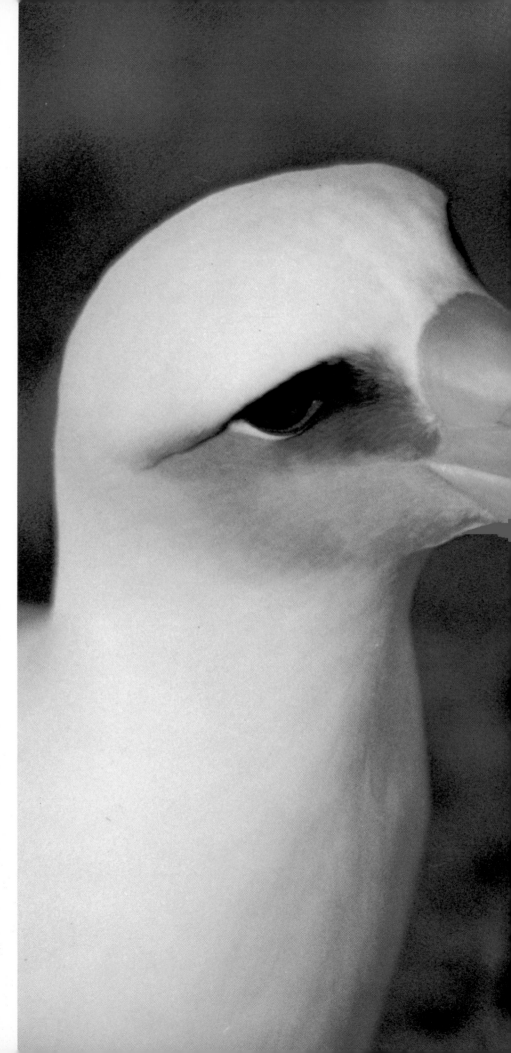

Is that a love peck one Laysan albatross (right) is giving another? Yes, they are enacting a mating ritual, but grooming each other's feathers serves another purpose. Preening is as essential to their survival as eating and all birds do it several times a day. At the end of the breeding season, the albatross must have clean, well-oiled feathers to fly for months over the Pacific, resting and feeding on the ocean's heaving waves. Like the whooper swan (below), the seabird rubs it beak on a large oil gland near the tail and distributes the oil through all the body feathers, removing parasites and dirt as it goes.

Such a lovely photograph of a swan performing this homely chore is no accident. Teiji Saga has dedicated his entire photographic career to taking pictures of the whoopers who winter in his native Japan. This is one of 20,000 swan photographs in his collection.

With rib-wrenching urgency, a barren grounds caribou (left) makes a desperate attempt to scratch an intolerable itch. Warble flies lay eggs on its flanks and the hatching larvae burrow into its skin. More deliberately, an Alaskan brown bear (right) dislodges a pest from its ear without interrupting its salmon watch on the McNeil River. Writhing and rolling in a Black Hills dust bath, a bison bull (below) turns its world upside down to rid its hide of ticks, flies, and mosquitoes.

To escape these unseen tormentors, helpless giants may swat and stamp, huddle together, seek windy hilltops, whack their way through underbrush, or even stampede; but the only way to spell relief from this universal summer scourge is w-i-n-t-e-r.

An itch has suddenly claimed the full attention of an adolescent male elephant (right) and he has decided to deal with it with the aid of a conveniently located rock. Closer to the waterhole, he might have taken a plunge or sought soothing relief in a mud bath. At other times, he lines up with his peers to take turns using a tree trunk or a large termite mound as a scratching post. A supple sea lion (below) also appreciates the businesslike hardness of rock surfaces for scraping off irritating parasites that can't be reached by a pretzel-like twist uniting head and rear flipper.

Even the white-footed mouse (left) must fight off mites and other free riders, so it gets the jump on them by meticulous daily grooming. It licks each toe, scratches its back, and combs the fur on its head. Lastly, the tiny rodent cleans its three-inch tail in this rarely seen performance.

A
Wildlife
Family Album

The Quest for Food

*O*n elfin wings a living jewel floats before a beckoning blossom. To human eyes, it's a picture of sheer poetry. But to the long-tailed sylph hummingbird (right), looting a flower's honeypot is sheer hard work. Those tiny wings must beat about 50 strokes a second to hold the feathered mite in midair while its fringed tongue sips the sugary nectar. And to keep its wings whirring, the sylph must feed hundreds of times a day.

We might call that a vicious circle: the bird beats its wings that fast in order to exploit a high-energy food source—which it needs because its wings beat so fast. But naturalists see a creature in perfect accord with its chosen fraction of nature's endless smorgasbord. Let the red-tailed hawk wrestle the rattlesnake, the pelican plunge for the darting fish, the buzzard peck at the carcass. The iridescent little hummer is content to take all its meals from the hearts of flowers.

In nature there is food in almost everything—plant, animal, the air, the water, the soil, the light of the sun. Even as nature spread this universal renewable feast, she devised ways to keep it in balance.

Lest the world be overrun with ants, anteaters evolved tube-like snouts and long, sticky tongues to thrust into anthills. Antlion larvae build tiny sand traps where they can suck dry any ant that wanders in. Bears too slurp ants along with their blueberries, and flickers pluck them from tree bark.

Because waters abound with fish, kingfishers and fishhawks discovered they could nab those that venture too near the surface. The fishing spider found it could climb down an aquatic weed stem and seize the unwary minnow. Bears and wolves paw the spawning salmon, raccoons the skittering crayfish.

Although nature provides water, not all creatures slake their thirst by drinking. In the Namibian seaside desert, beetles rear their nether ends to absorb moisture from nightly fogs. In the American Southwest, kangaroo rats synthesize water from carbohydrates in the seeds they eat.

Getting food and water is often dangerous. An animal may have to expose itself to an enemy or even to a pre-emptive strike from its own intended meal. But that is a risk that must be taken as hunger and thirst propel each creature through a life-long search for its next—and possibly its last—meal.

(Overleaf) The distance between the Alaskan brown bears in this photo tells you the McNeil River was full of salmon the day the picture was taken. When salmon are scarce and these pugnacious bears are ravenous from a meager spring diet, fights erupt often for the best fishing spots. A bear catches a slippery salmon by pinning it to the bottom with its claws, then grasping the fish in its jaws.

102

The powerful cheetah (left) nonchalantly lapping up a drink from this African waterhole can take its time. Few animals would dare attack. Cheetahs can go for days without drinking, even in hot weather, perhaps because they consume the blood when they eat their prey. It's not unusual for a gerenuk (right) who lives in arid bush country to do without water altogether, although individuals living near water may imbibe occasionally. The non-drinkers get moisture from succulent acacia leaves, plucked from between the thorns with their tongue and upper lip, often at heights of up to six feet.

(Overleaf) South African impalas need water at least once a day and often put their lives on the line to get it. The timid animals time their visit to the water hole so that they can drink in peace while their predators take their mid-day nap. Even then, at the first hint of danger they will bound away with 30-foot leaps without drinking a drop.

The golden eagle (left) swoops down at 120 to 150 mph to catch a fast-escaping rabbit, prairie dog, or squirrel on the ground or a goose or crane on the wing. The lion's technique is to slink through high grasses, just like a domestic cat, to get as near as possible before being detected. Yards before reaching a wildebeest or other prey, the lion pounces from its ambush with a final 35 mph dash. Usually females do most of the tracking and killing, but the males always get the lion's share (below). Pushing lionesses and cubs aside, they gorge themselves on the kill.

Nuts and berries are favorite foods of forest dwellers. The chipmunk (far right) hides away nuts and seeds for winter eating before other small mammals, game birds, or deer beat it to the windfall. The common box turtle (below) is so attracted to blackberries that it may lean on a rock to balance itself on its hind legs and the edge of its shell to grasp a high-hanging berry. A black bear (right) will raid a camp garbage can or rob the chippie's acorn cache and eat the chipmunk as well. In fact, this nonstop snacker eats anything from fruit and wild honey to skunk cabbage and carrion.

*F*ishing is easy work for animals made for the job. The sea otter (above) uses its sensitive forepaws to find seafood in a kelp bed or on the sea floor. Then with rounded teeth and strong jaw muscles and using a rock as a tool, the otter cracks the shells to get at the meat. The Florida alligator (far right) simply followed a fisherman's line and picked up this discarded mudfish. But an alligator usually fishes under water. Transparent eyelids protect its eyes, and valves in its nose and throat keep water out as it dives open-mouthed after prey.

The shape of a puffin's mouth (right) helps it to collect enough fish in one haul to feed its waiting chicks. It catches one fish at a time, then with its tongue files each one in backward-slanting serrations on the roof of its mouth.

The bald eagle seldom wanders far from water and the reason is clear: fish, dead or alive, are the staple of its diet. Less often, eagles also catch muskrats, rabbits, squirrels, and injured waterfowl. To catch a fish, an eagle scans the shallow water of small creeks with binocular vision, then zooms in feet first on the target and grabs it with its talons. Two "baldies" (right) in the Karl Mundt Wildlife Refuge in South Dakota chose a favorite cottonwood perch to feast on their Missouri River catch: fresh crappie and walleye pike.

(Overleaf) Better luck next time for this bald eagle! Swooping down on the Snake River, it grabbed too large a fish—one which proved more than equal to the challenge. The fish jerked the startled eagle under the water several times until the humiliated "imperial ruler of the sky," out of its element, let go. The soggy bird then floated back to shore, shook itself, and stretched out its wings to dry.

The feeding ritual of the Siberian white crane (left) and the Eurasian whooper swan (below) may remind you of a dance — the graceful ballet of the tutu-feathered cranes or the comic bottoms-up cancan of the swans. In their shallow marsh, the cranes move majestically to and fro, reaching for submerged animals and plants with a notched beak made for tugging roots and reeds. In deeper water, the whooper swans feast with tails up and feet fluttering. Their fast footwork agitates the water, loosening plants from their muddy moorings. Both species escape arctic ice and snow in these milder winter feeding grounds.

119

$\mathcal{Y}$ou can't argue taste—even in the animal world. Both the wild boar (above) and the raccoon (right) have an eclectic palate. The sow and her two-to-three-week-old piglets are foraging for nuts at the foot of a beech tree in Germany. Wild boar are mainly vegetarian, digging with their tusks for roots, or looking for grains and plant stems. But they often eat insect larvae and, on occasion, even carrion. Raccoons change menus with the seasons. In spring and early summer, they eat protein: frogs, snakes, turtle, mice, or birds. The rest of the year they feed on fruit, seed, or grain.

The long-nosed bat (far right) is much less adventuresome when it comes to eating. It is perfectly content with a steady diet of nectar from the century plant. Its elongated, fringed tongue enables it to pick up all the high-energy nectar and nutritious pollen it needs.

To bees, butterflies, and some birds, a flowering meadow is a spa with free-flowing ambrosia. They come, they drink, and thank the host by leaving life-renewing pollen gathered at their last "pub stop." As the honeybee (left) fills its honey bag with nectar, it wiggles its fuzzy body in a bed of pollen dust. It combs some into hairy cases on its legs, and bears the rest to the next dandelion. In a similar way, the fritillary butterfly (far left) picks up and delivers sticky packets of milkweed pollen, and the forktailed royal woodnymph (below) collects and deposits thistle pollen.

A Wildlife Family Album

The Need for Shelter

_S_unny skies blacken into rain clouds. Day darkens into night. Warmth gives way to cold as summer surrenders to winter. April's torrent is forgotten in August's drought. A tranquil scene turns ominous with the sudden presence of a predator. One of nature's constants is its inconstancy; minute by minute and era by era, a wild animal's surroundings change. For change measured in millenniums, the answer is adaptation; those creatures without the answer face extinction. For change measured in moments, the answer for many is shelter. A place to hide, to rear young, to stay dry, to keep warm, to sleep. A den, a burrow, a nest. A home.

Tiny ogre in a sultan's palace, a jumping spider (right) peers from the elegant arch of a Turk's-cap lily. Few of its enemies will notice it in hiding—and few of its victims will notice it in ambush. For this eight-eyed little huntsman, shelter is for both hiding and seeking. The lily was there for the taking. So is the empty shell that the hermit crab backs into, and the bigger one it backs into later when it has outgrown the first. But most birds, and many insects are makers rather than takers of shelter. One of the more elaborate builders is the rufous-breasted spinetail, a Central American ovenbird whose chambered and tunneled retreats have earned it the nickname of the castlebuilder. Inside its three-foot-long untidy collection of twigs lodged in a tree or shrub, snakeskin fragments carpet a long hall leading to the nest chamber. There the eggs are laid on a bed of fresh green leaves.

The gall wasps have some gall indeed; they make plants build their home. Egg-laying forces a stem to form a hard green ball around the eggs, giving the larvae not only a cozy hideaway, but a built-in food supply. The dormouse is one of the few mammals that construct an intricate shelter—a woven, roofed-over nest attached to a swaying low branch. Chopped grass and soft plant fibers line the cradle. A few reptiles neither take nor make—they are their own shelter. The box turtle comes fitted with trap doors fore and aft; who can breach such a fortress? The common chigger and the wood tick can. But a sniffing, pawing red fox will soon give up and look for more vulnerable prey to carry to the den he and his vixen have dug. Then the turtle sticks its feet and head out the doors and shambles away, mobile home and all.

(Overleaf) It's only a bare spot in the grass, but this nest has what skittish arctic loons consider important: an emergency exit. The returning parent (lower center) climbs up a short muddy ramp that connects nest and lake. At the first sign of danger, the loons flatten their bodies. If the threat persists, the whole family belly-slides into the water, submerges silently, and surfaces far from the nest.

126

Instead of digging a den or a burrow as the red fox does, the gray fox (left) often takes shelter in logs as this sleepy pair has done. The secretive little opportunists may also den in rock piles or crevices, in abandoned barrels or buildings, or under a barn or chicken house. They may line the lair with leaves, grass, bits of bark, or fur. Because the gray fox climbs trees, an occasional den may even have a penthouse view.

You might come across a charming sight like this one in your community if you were to inspect every hollow log and culvert. Though seldom seen because they hunt at night and sleep all day, gray foxes thrive in the brushy or wooded areas of nearly all the lower 48 states. If dawn catches them far from home, they curl up in a new shelter and sleep 'til sundown.

A ledge is only one of many building sites that the white-footed mouse (below) uses, but the bald ibis (right) will rarely settle for anything else. The mouse's favorite place is in trees, the taller the better. But in empty birdhouse, beehive, or wherever, this nonstop breeder weaves grasses into a comfortable home for each of its several litters per year.

The bizarre-looking ibis makes do with a very skimpy nest, but cleans the old site before building anew for its annual brood. Tragically, this goose-sized, old world bird, which once graced the groaning board in medieval castles, may not return for many more spring housecleanings on this cliff above the Euphrates River in Turkey. This surviving remnant flock is now down to about a dozen pairs. A few small flocks are also hanging on in Morocco, but the outlook for the dwindling species is bleak.

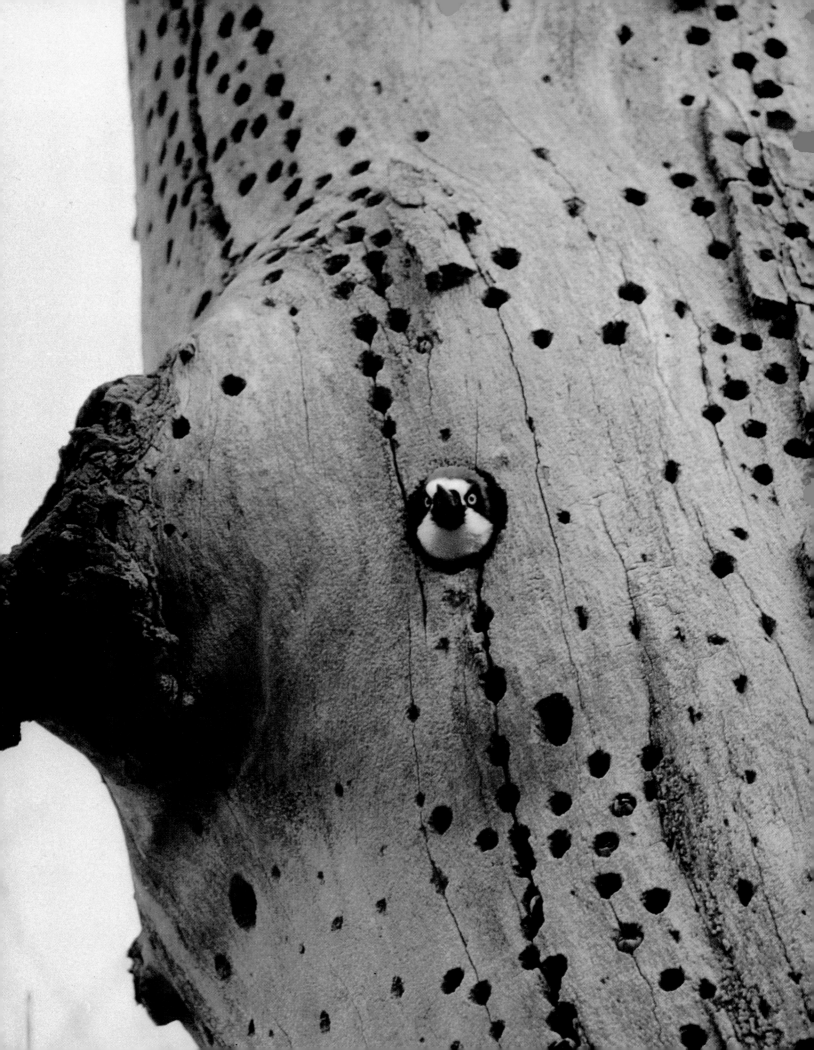

*A*corn woodpeckers excavate cavity nests two feet deep inside this old oak (far left) and drill food storage holes on the outside. They stock the holes each year, driving acorns in tightly so raiders can't snitch one before the owners pop out and zap them.

If the osprey (left) touches two power lines, it's dead, and electrical service is interrupted. To try to help these birds, utility firms build more appealing nest platforms away from the wires.

Who can top the aquatic spider (below)? It lives in an underwater web inflated by air it brings down, one bubble at a time.

In a mossy log a young porcupine (left) has found refuge from the cold rains and snows of a northern forest, while a banner-tailed kangaroo rat (below) scurries down its burrow to escape the heat of a southwestern mesa. The porcupine will have three types of hair and some 30,000 quills by the time it grows up; yet in spite of its unique armor, the "quill pig" still seeks protection in a log, cave, empty building, or borrowed den. But if deep snow buries these shelters and the plants it eats, the well-insulated porky may climb a tree and stay there 'til spring, living off the bark.

The kangaroo rat's underground fortress includes a pantry for hoarding seeds and a grass-lined nursery for its next litter. To keep heat out, the bannertail fills the real entrance with sand. It also digs false entrances that lead to dead ends, stalling the desert fox's attempts to pay a call.

A female red-shouldered hawk (right) snaps off a cypress twig to carry to her nest in Everglades National Park (below). Ornithologists aren't sure what this gesture means, but the results appear to be both decorative and good for her brood. Before the eggs hatch, the touch of greenery shades them from the Florida sun. Later, foliage gives the fuzzball chicks a cleaner, cooler place to live. For 60 days the female keeps freshening up the old stick nest in this way until the young leave for good. Meanwhile, the male brings her food—a young black swamp snake, a frog, a mouse—easily found in the swamp.

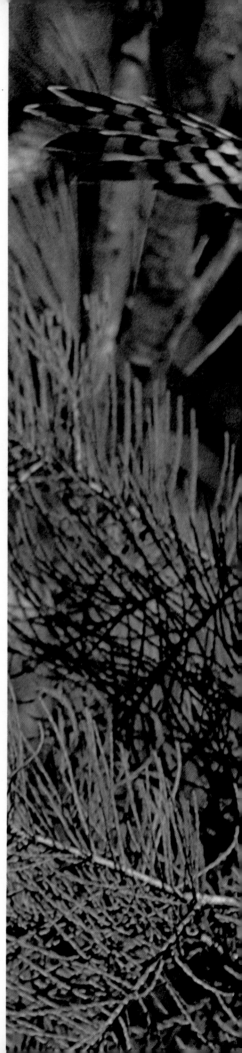

This hawk must build where it can live on small animals and where there is less competition with other raptors that have larger, stronger talons for hunting in more open country. In the woods, the red-shouldered hawk has found its niche, one of the few species whose choice of nest site is influenced by its feet.

This sunset silhouette (left) dramatizes the grand scale on which great blue herons live. The regal four-foot-tall birds spend much time like this standing in their massive stick nests that are three to four feet across. Sometimes 30 or 40 pairs build in the crown of one large tree.

The nests in this Ohio roost appear to be old ones that have been enlarged each year. A first-year nest is only a flimsy platform which a new pair has put together, he gathering and she arranging, with much ceremonial twig-passing. If more than one heron species shares the tree, the great blues usually claim the top story.

Half a world away, in the green gloom of Australia's east coast jungle, a six-inch rufous fantail (right) has crafted an exquisite nest in the shape of a wine glass two inches across and one inch deep. Spider's silk binds its dried grasses and bark fibers and soft, fine rootlets line the cradle for the nestlings.

A Wildlife Family Album

The Mating Season

*F*ighting. Dancing. Strutting. Bullying. Gift-giving. What manner *of behavior is this from animals whose lives proceed through most of the year with no such theatrics? We might call it spring fever, but on nature's agenda—be it spring for some, fall or any old time for others— it is time to mate.*

Suddenly the rules of order bend. Bull elephant seals (right) that swam the seas in peace now turn their breeding beaches into battle- grounds. A few months before, they sprawled placidly on the sands, shedding fur and a layer of skin with a neighborly decorum that lasted for weeks. But now there's more at stake; now the bulls bloody each other and snort through their bulbous echo-chamber snouts in epic duels over the waiting females. For this largest of seals, a clash between bulls 15 feet long and two tons heavy can be a bout of titans indeed.

The males of many a species must start the quest for females by dealing first with other males. Desert tortoises may lock in a slow-motion shoving match that leaves one victorious and the other upside down. Unless the loser rights himself before the sun turns his shell into an oven, he loses not only the mating rights but life itself. Zebras clash in spectacular fights to determine which will be the family stallion. Two males circle one other, each trying to bite the other's legs. To protect their legs, they drop to their knees, still circling. Suddenly they may leap up, stand on their hind legs, and strike and bite at each other until one has had enough. The loser leaves and the winner claims the mares.

Sometimes male animals fight for land rights and the females come with the territory; sometimes the suitors fight for mating rights and worry over territory later, if at all. Either way, the species makes sure that only its strongest will sire the coming generation.

Often, all the male contestant wins is the right to court the female; she will decide whether he wins the right to mate. Is he the correct species? By coloration and ritual he may satisfy her that he is. Will he be a good provider? His gifts of food and nesting material betoken his commitment. Is he ready to mate—and to excite her to readiness too? See him strut and display his finery; watch him prance and gyrate and stroke and nuzzle; hear him flute and bugle and drum. 'Tis springtime, and what female can resist?

(Overleaf) The male Australian red- capped robin doesn't let his colorful good looks tempt him to stray from his drab mate. From the moment she begins to respond with an agitated fluttering of wings to his courtship gifts of food, the couple become strongly bonded. After mating, he feeds the female while she incubates her eggs. From July to December the male may help her to rear four broods.

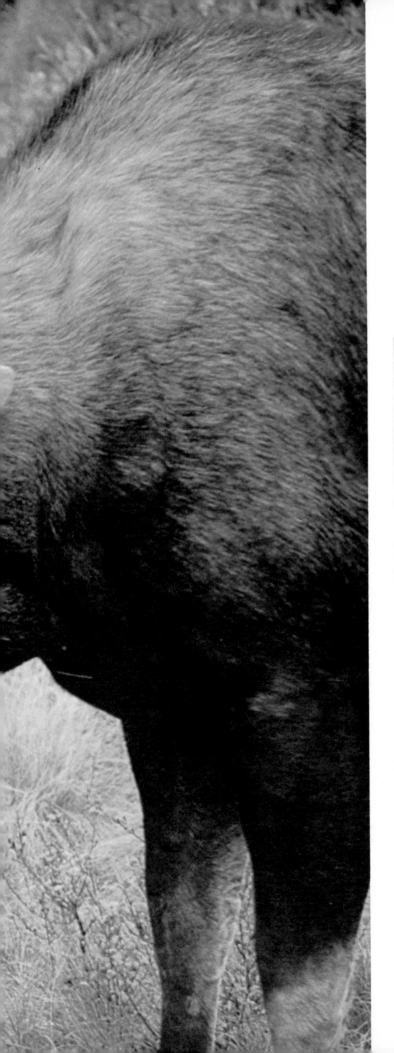

When two rivals eye the same female, jealousy flares. Rutting bull moose (left) show belligerent feelings by bristling up their manes and lowering their ears before charging. Most encounters produce more bluffing than sparring, but sometimes antlers lock and the imprisoned bulls starve. The winner mates with the cow several times a day until she refuses any more of his attentions. Then the bull goes off in search of a new mate. A furious Canada goose (below), letting all his anger explode in a flurry of wings, blusters after an aggressive rival who was flirting with his intended mate.

(Overleaf) A late spring encounter in the forest between two male grizzlies turns into a savage fight. During mating season, grizzlies vying for territory acquire nasty tempers. Many an older bear shows battle scars from countless rivals' teeth and slashing 4-inch foreclaws.

145

On a courtship run, salmon hasten to the spawning site in the cool waters of an up-river shallow stream. Pre-spawning changes cause sockeyes' silvery skin to turn red and olive green and the males' jaws (above) to form a long hooked beak which they use advantageously to fend off rivals. A persistent intruder (right) gets a sharp bite on the neck. The male (far right) stands guard while his mate excavates a nest in the gravel bed with her tail. After the male fertilizes the eggs, the female brushes pebbles over them. This may protect them from predators and swift currents until they hatch.

(Overleaf) With neither bellow nor bite, two hippos gently touch jaw to jaw in a courtship gesture before mating under water. Nevertheless, the chauvinistic male completely dominates the female. Should she resist his advances, he may yet resort to biting her into submission.

149

*T*hree anxious suitors seem to be performing a graceful water ballet around an algae-covered female manatee off the Florida coast. When the males discover that the female is in heat, they nuzzle her and try to maneuver into a mating position. For two to four weeks, she manages to elude them by swimming away until she is ready to mate. Finally, she stops playing hard-to-get and mates with one or more of the 500-to-1,000-pound males.

Unlike most animals, male manatee groups rarely show competitive aggressiveness when trying to win a female. When the game is over, the herd disperses until the next cycle starts. Manatees mate any time of year and the female bears her young underwater about 13 months later.

*E*n garde! A male yellow-billed stork (right) challenges his rival for a nesting site. Their long slender beaks, necks, and legs, and their fluttering wings give a patrician elegance to their bill-clattering duel. To court and win his chosen mate, a red-footed booby (below) offers the female nesting material which he stole from unoccupied nests in his colony in the National Wildlife Refuge on Tern Island, Hawaii. With a raucous squawk, the male lands on her back and presents his gift of straw before copulating. The male makes repeated forays for additional offerings during the two-to-three-day mating season.

ife begins with a dance for the regal six-foot Japanese red-crested cranes (right). To trigger hormonal changes in the female that make fertility possible, the male begins by approaching his prospective mate with a stately bow (top). The pair then interact with a series of forward steps and hops (center). A fluttering, twirling black-and-white finale climaxes the courtship ritual (bottom).

To boost the population of the endangered whooping crane, an American ornithologist (left) joins in a similar dance at the International Crane Foundation in Wisconsin. This dedicated scientist was hoping to prepare a lone captive female for artificial insemination and it worked. After dancing with her strange "mate" for two to three minutes, the responsive female solicited copulation. The scientist stroked her back, then injected semen from the Foundation's bank.

A Wildlife Family Album

Threat and Rescue

*T*houghtlessly, we kill. Thoughtfully, we save. What a paradox is this two-legged animal who paves and poisons the world that he too must live in, and then spends his energy, and occasionally even his life, to rescue creatures in peril from demons that he himself may have loosed. A howler monkey plucked from a dam-flooded jungle valley (right) squirms and screams in unknowing terror of the Panamanian rescue worker who gives the animal back its life. Yet the numbers of wildlife evicted by flood waters are few compared to those whose homes have been destroyed by polluted water, bulldozer, and plow. But that is the real world that man has made, and every animal must try to survive in it. Wildlife's album must picture man as well.

Happily, some of the scenes are heartwarmers. In them we watch our kind risking bites and clawings, braving steamy heat and arctic cold, spending time and dollars and energy and brainpower, sometimes to rescue, but more often to study how animals live. Indeed, if we are to prevent future crises, we must learn everything we can about the lives of our wildlife family. Otherwise we may discover too late that in ignorance we have imperiled a creature or a whole ecosystem by an act we thought would be harmless. Had we known long ago that the sulphurous oxide in factory smoke can travel hundreds of miles on air currents to fall as acid rain, we might have put scrubbers in our smokestacks sooner. Because of our ignorance, some Adirondack lakes once leaping with fish now languish like liquid deserts devoid of life.

Now that we know that eggshells get thinner and break in brooding as certain pesticides work their way from bug to bird, we have banished those sprays from our skies. Now we say nay to plume and trophy hunters, to shoes of alligator hide and to grisly umbrella stands made of elephant feet. And we give of our money to join with those who give of their time to weigh the hibernating bear or mend the splintered wing or bathe away the crippling scum of oil. Our reward comes when the evening news reports that plans for a dam have been revised to protect a rest area for migrating whooping cranes and that men from every major nation are conferring over our dwindling stocks of whales. Then the world becomes more secure. For every animal. Even the two-legged paradox.

(Overleaf) When men and animals compete for the same resource, both ultimately lose. Claiming dolphins eat commercially valuable fish, Japanese fishermen kill them by the thousands. Environmentalists urge fishermen everywhere to try other solutions: acoustical gear to chase dolphins from fishing grounds or government subsidies for a reduced catch to allow overharvested fisheries to recover.

The outboard manned by these Choco Indians became a modern ark for drowning or starving animals caught in a man-made flood in Panama in 1976. Rising waters spread into the jungle behind the Bayano Hydroelectric Dam, forcing deer to swim and small mammals to climb trees. Rescuers had to top the tree where this terrified coatimundi (right) clung. They rushed it to a rescue camp for first aid and care before releasing it to the wild. The baby black howler monkey (below and page 161) regained strength on a special formula. Operation Noah II saved 3,641 animals.

*H*elping hands may make a difference in the future of these exotic infant gharials from India. From the embryo stage to adulthood, these relatives of the crocodile lead a precarious life. Poachers steal eggs, which are a gourmet's delicacy. Predators—mice, wading birds, fish, small mammals—prey on hatchlings. Adult gharials die from explosives illegally used by fishermen or when they are taken for their valuable hides. New dams flood, and pollution spoils, portions of their Chambal River habitat.

Local and international agencies have stepped in to keep the endangered species alive and well. These caretakers remove eggs to protected hatcheries and rear the young for three years. As the juveniles come of age, the workers release the three-to-five-foot gharials to sanctuaries where no commercial fishing is allowed. The nearly 2,000 young now being raised will improve the survival chances of this species, which is down to less than 200 adults in the wild.

165

$\mathcal{A}$ friend in need, 18-year-old Alberto Palleroni willingly accepted the responsibility of caring for an injured bald eagle. A state policeman brought the crippled bird to Alberto, who is well-known in Eugene, Oregon for his raptor expertise. X-rays showed a fractured wing bone with infection. Despite the aid of a veterinarian and an orthopedic surgeon, and two years of Alberto's good care, this eagle did not survive. Alberto often nurses his charges up to three years, teaching the big birds to fly again on a 40-foot tether before releasing them to the wild.

This young man is one of a growing cadre of knowledgeable volunteers who care for injured or orphaned animals. Although non-professionals themselves, they offer warnings to would-be rescuers who are not schooled in wildlife care: young animals usually are not abandoned —mother is probably nearby; a sick animal needs constant attention from a qualified person; anyone holding a protected creature needs a federal permit; the rate of survival is generally low in most cases of amateur wildlife care.

*I*t takes imagination to realize that the oil-soaked western grebe (right) was once as beautiful as our incubating eared grebe on pages 6 and 7. Caught in the Santa Barbara oil spill in 1969, this grebe was among thousands of birds that died. The 4 to 6 percent survival rate of ten years ago has now risen to 60 to 70 percent as rescue teams have learned how to clean oil-covered birds. The South African black-footed penguin (below) was one of the lucky ones. Rescuers repeatedly dust the feathers with powdered clay to absorb the oil, then scrub the bird in a grease-cutting detergent bath.

*Y*es, you have no bananas? the inquiring chimpanzee seems to ask as it peeks under Dr. Jane Goodall's shirt—her usual hiding place (immediate right). Goodall and photographer Baron Hugo van Lawick kept a store of bananas in strongboxes (lower right) to reward cooperation during their studies in Tanzania. Similarities between chimp and man—in biological makeup and the ability to share, to make affectionate attachments, and to communicate non-verbally—make these studies an invaluable tool in the search for the roots of human behavior.

The cheetah (far right) hitching a ride on zoologist George Frame's land cruiser is one of about 1,000 living in the Serengeti National Park. Frame's four-year study found them doing well here where regulations prohibit hunting and restrict tourist vehicles, and managers control the burning of dead grasses to give cheetahs cover for stalking and for concealment of cubs. But elsewhere, hunting and the spread of farming and forestry are diminishing habitat for both chimps and cheetahs.

(Overleaf) Dian Fossey's 14-year study of mountain gorillas shows they can express abstract, even speculative, ideas in sign language. Future research hangs on increased protection of this vanishing species. There are only 220 left in the world, all in Zaire, Uganda, and Rwanda—down 50 percent in the past 20 years.

170

A helicopter and a dart gun loaded with tranquilizers enabled scientists to get close enough to wild polar bears to conduct this 1972 survey of these former monarchs of the arctic wilderness. Researchers from several nations worked together to weigh the huge animals (right) and to conduct physical examinations (below). The biologists wanted to find out how the bears could be protected from overhunting and from the oil and gas drilling, mining, new roads, towns, airports, and the shipping lanes that were encroaching on the 100-to-300-square-mile territory needed by each bear. In addition, potential oil spills would harm ringed seals, the bears' food staple.

Thanks to their work, Russia, Canada, Norway, Denmark, and the U.S. agreed in 1973 to allow only natives to hunt and to set up programs that now improve the bears' chances in a changing environment.

Bear studies have boomed in the U.S. in the last ten years as growing numbers of wilderness hikers invade the bears' turf. By attaching radio-transmitter collars to the adult bears (far right), biologists track their wanderings and recommend to park planners the rerouting of hikers' trails to safer areas. Counting black bear cubs (far right) and weighing the adults (right), then relating gains and losses to food supply helps officials determine hunting quotas in overpopulated habitats. A health check-up (below) of this sedated grizzly may uncover treatable disease, infection, or parasites.

176

The Dalmatian pelican (left) and the European beaver (below left) in Russia and the puffin (below right) in the U.S. are making a comeback. This pelican almost became extinct in 1920, increased to 300 pairs by 1949, and is now thriving in the rich, wild marshlands of the 153,000-acre Astrakhan preserve. Soviet scientists have also restored the beaver from a low of 1,000 in 1934 to 150,000 today.

While other puffin species have not been endangered, excessive hunting wiped out all but one U.S. Atlantic puffin colony about 1900. Since 1973, the Canadian Wildlife Service and the National Audubon Society have been transplanting 100 chicks a year from Canada to a new colony off the Maine coast. Biologists hand-rear the chicks for a month, then band them before releasing them. The bands serve as I.D. bracelets for migration, population, breeding, and nesting studies.

Work was a wrestling match for these two University of California biologists when they tagged seal pups like this frisky 400-pounder in 1977. The men were studying breeding and migratory habits of the northern elephant seal on San Miguel Island in the Santa Barbara channel.

These marine mammals became endangered when 19th century sealers harvested them extensively for their fine-quality oil. In 1911 when only a few seals were left on Guadalupe Island off Baja California, a Mexican ban on hunting turned the tide. Left alone, the seals flourished as their hierarchy system that allows only the strongest to mate produced a hardy stock. By 1976 the estimated population along the Pacific coast from Alaska to Mexico had soared to 50,000.

181

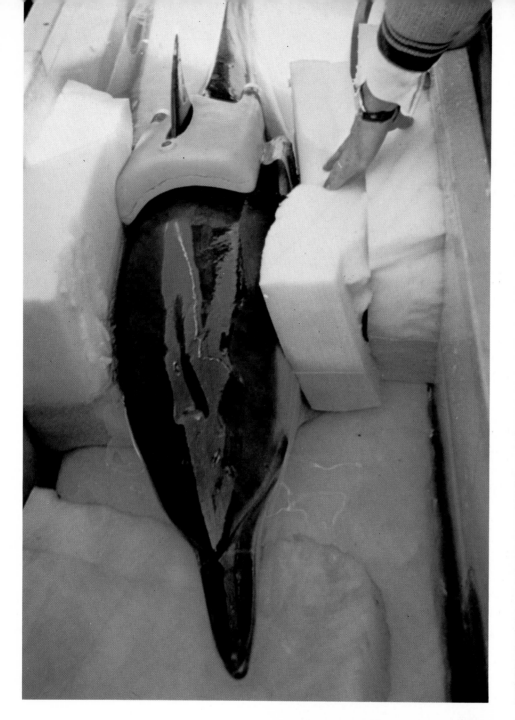
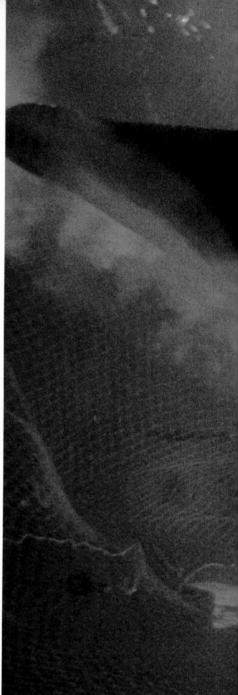

 Dolphins—like all creatures—act naturally only when at liberty. So the Jacques-Yves Cousteau research team works where the dolphins are—at sea. To determine whether dolphins use vision to navigate, they covered the animal's eyes with foam rubber blinkers (lower left). Deprived of sight, the dolphin maneuvered without touching barriers, proving that dolphins can navigate solely with their super-efficient sonar system.

In a tracking test, the biologists attached a radio-transmitter to the animal's dorsal fin (upper left), but the cumbersome device kept the dolphin from catching up with the school. In another try, a diver (above) marked a dolphin with a harmless fluorescent chemical. The dolphin disappeared in the deep, the chemical diffused, and the dedicated researchers—always learning even from "failed experiments"—used their knowledge to search for a better way.

Who would risk their lives to rescue a shark? Researchers Ron and Valerie Taylor would. When this giant white shark tried to break into the Taylors' cage during a filming session off the coast of Australia, it became hopelessly entangled in the rigging (left and below). The Taylors bravely left the safety of the cage to free their underwater photographic subject. But before releasing the shark, they tagged it and took its temperature.

To this husband-wife team, every white shark counts. The scarce giants have a low reproductive rate. At the same time, sharks now rank as popular big-game fish, so sport fishermen, aided by advanced fishing gear, are catching them in increasingly large numbers. Data on the sharks' needs and habits collected by investigators such as the Taylors give regional fisheries planners a sound basis for regulating fishing to avoid over-exploitation.

A
Wildlife
Family Album

The Circle Closes

*J*ourney's end. The spark of life flickers out. It is, after all, only a borrowed spark, and sooner or later nature has to recall the loan.

Whatever we humans share with the other creatures of earth, we alone know that each life must end. Around that shared awareness clusters much of our philosophy, our religion, our charity—much that makes us noble, gives us wisdom, teaches us humility and thankfulness.

Thus do we marvel at the ant that joins with others to form a bridge and, while its comrades cross the streamlet, drowns. Thus can we recognize nature's wisdom in strewing the pond with countless frogs' eggs so a few may mature as the many feed the hungry. Thus can we accept the chilling efficiency of the python even as we regret the loss of the gasping rodent. And thus can we take our share from nature's treasury, whether as Africans hunting the antelope (right) or as Americans pushing the shopping cart—not with guilt but with gratitude.

For every creature, death is life's bottom line. But for the ongoing miracle of life itself, there is no bottom line. The dinosaur died out a million years ago, yet lives in its evolutionary descendant, the soaring bird. Many a male praying mantis dies even as he mates, yet lives in his offspring as the female devours him to nourish her eggs. The fasting female octopus starves to death as she guards her eggs; yet she lives on in her hatchlings.

But isn't it unfair, we start to ask as images of triumph for one creature and tragedy for another fill our album's final pages. The victor gains a brief reprieve from hunger, but the vanquished loses all. Yet our protest reveals our forgetfulness of the ancient relationship between predator and prey. That relationship is, after all, the foundation on which all the planet's ecosystems rest.

Death is part of the design which ensured that we would be born into a world alive with the beauty and the excitement of wildlife. As we grow in our understanding of that world, it is the wholeness of nature that wins our respect. Then we see not a world of winners and losers but of living creatures, each linked to all the others that came before and those that will follow. For each life, the circle one day closes. For all life, the circle never ends.

(Overleaf) Life ends for an elk in the snow of a Canadian forest, and life is sustained for five timber wolves—until hunger drives them to hunt again. This familiar drama goes back millions of years to evolution's edict that some animals would eat only plants, some only flesh, and a few would eat both. That decree holds all three groups in check and explains why so few wild animals die of old age.

The sight of a once lithe, graceful gazelle transformed into meat for a hungry leopard makes a powerful statement about the ways of the wild. Yet nature prepares even its gentle creatures well. Over eons, it gave the gazelle all the tools of escape: alertness, a fine sense of smell, a 40 mph sustained sprint, and an innate fear of thickets where enemies stalk.

Even for the swift leopard, capturing a gazelle is unpredictable. All circumstances have to be right. Observers of the kill would have glimpsed the big cat stealthily approaching in the tall grasses bordering the open field. Counting on its overwhelming power at close range, the leopard made a lightning rush toward the herd, seizing the grazer that was least physically able to escape. The victim may already have been weakened by disease, injury, or old age. If so, its death actually helped its kind by leaving behind on the African savannah a stronger herd to carry on the species.

A man and his dogs (below) search a frosty Nebraska field for signs of pheasant or quail for the holiday table. For both hunter and dogs this dawn is a homecoming, a renewal of their shared bond with the world of nature.

The fisherman (right) may make a hearty chowder from one of the codfish at his feet, but most of his catch will be eaten by others. More likely he measures this day's work in the Grand Banks off Greenland in terms of payments on his seaside cottage. And he worries that modern factory ships are rapidly crowding out his centuries-old handline. On a good day a doryman once caught a ton of cod in these waters.

Reluctant to sever for all time man's first role of natural predator, sportsman and old-style fisherman alike prefer to match wits with their prey. They choose the challenge and the immediacy of the stalk or the lure —and to gamble on the feast.

*N*o conservationist dilemma deters this Ugandan (above) from taking a precious ivory tusk for income to buy life's necessities. Brought up in a remote, finders-keepers world, bush people may not even know that the elephant is threatened throughout its entire range. A young New Guinean (right) too was taught that any wild creature is fair game. In his native tongue in Bulolengabip, the word for meat and enemy is the same.

Elsewhere in New Guinea, (far right), a new idea is taking hold. To help save the once-vanishing crocodile, the government has set up a rearing program in which young wild hatchlings are raised for the commercially lucrative skin trade. Raising these young crocodiles in pens rather than hunting adults leaves the breeding stock to flourish in the wild. A Papuan villager inspects a hide that is the correct size for sale in the international market.

*D*eath comes in bizarre ways to insects that blunder into insect-eating greenery like the pitcher plant (left). Lured by its bright color and sweet nectar, a leaf-hopper skids on the plant's waxy inner wall and falls into the pool at the bottom. Downward-pointing hairs frustrate the insect's efforts to escape as enzymes and bacteria in the water begin to digest it.

Unaware that it faces a similiar fate, a coppery-eyed lacewing (below) strolls onto a Venus' fly-trap. When it accidentally touches trigger hairs, the two leaf blades snap shut (bottom). Entombed, the lacewing will be absorbed in three to five days.

197

Whenever a whale runs aground on the East Coast, Dr. James G. Mead, associate curator of mammals at the Smithsonian Institution (left), will almost certainly arrive shortly to find out why. If the whale dies and isn't too big, he hauls it into Washington, D.C. (below) to study.

The difficulties of observing live whales keep Dr. Mead and other whale experts from learning if an epidemic, a berserk "leader," or other factor brings whales ashore as in a 1973 beaching of 35 pilot whales (far left). In shallow water, their echolocation system fails, they breathe in sand, and usually die of pneumonia.

Nature is not sentimental about baby animals. This little antelope (below) is about to lose a vital race with a cheetah cub that is learning to catch prey under its mother's supervision. Another cheetah mother (right) has to rest before she can eat the full-grown Thomson's gazelle she has just downed for her fast-growing hungry litter. Without the day-in-day-out, tireless hunting and constant vigilance of this veteran, her cubs might not have survived their perilous first three months —when over half of all cheetah cubs are taken by lions, leopards, and spotted hyenas to keep their own litters alive.

(Overleaf) Yet animal mothering has its lighter moments. An inexperienced common tern may have to try several times before she catches a fish tiny enough for her frantically hungry chick to swallow whole. And she will bring fish after fish until she finally gets one the right size. This parental instinct to feed the young is one of nature's most powerful ways of ensuring that the gift of life will always be passed along from generation to generation.

Index

Photo Credits

Pages 2-3: Teiji Saga. 4-5: Paul Chesley.

A Time to Be Born
Pages 6-7: Wayne Lynch. 9: Stephen J. Krasemann/DRK Photo. 10: Robert B. Smith. 11: Roger Tory Peterson. 12: Rajesh Bedi. 13: Clem Haagner. 14: James D. Young. 15: James Chambers. 16-17: Norman Myers/Bruce Coleman, Inc. 18-19: Stephen J. Krasemann/DRK Photo. 20-21: Rachel Lamoreux.

Growing Up in The Family
Pages 22-23: Simon Trevor. 25: Michael C. T. Smith. 26: Donna Rogers. 26-27: Lynn Rogers. 28: Hans Reinhard/Bruce Coleman, Inc. 29: Stephen J. Krasemann/ DRK Photo. 30-31: Jeffrey O. Foott. 32-33: Joseph S. Rychetnik. 34: Dr. William Weber. 34-35: Top: Hans Reinhard/Bruce Coleman, Inc.; bottom: J. Perley Fitzgerald. 36: Leonard Zorn. 37: David A. Caccia. 38-39: Dr. M. Philip Kahl. 40: Udo Hirsch. 41: Tim Fitzharris. 42: Ernest Wilkinson. 43: Allan Roberts. 44-45: Keith and Antje Gunnar. 46: Lester Peterson. 46-47: Hans and Judy Beste.

Leaving the Nest
Pages 48-49: Hal H. Harrison. 51: Thase Daniel. 52: Ron Austing. 53 Top: Jack Dermid; bottom: John L. Tveten. 54-55: Dan Davidson. 56 Bottom: Sam Blakesley/ Photri. 56-57 Top: Stephen Dalton/ N.H.P.A. 58-59: Wendell D. Metzen. 60-61: Roger Tory Peterson. 62-63: Dr. M. Philip Kahl/Verda, Intl. Photos. 64: Thase Daniel. 65: Robert Winslow.

Coming of Age
Pages 66-67: Tui De Roy Moore. 69: Elbridge Merrill. 70-71: Howard Hall. 72-73: Norman Myers/Bruce Coleman, Inc. 74-75: Wolfgang Bayer. 76: Bill McRae. 77: George Holton. 78-79: Jack Drafahl, Jr. 80-81: Steven C. Wilson. 82 Top: Kjell Sandved; bottom: Hector Rivarola. 83: A. van den Nieuwenhuizen. 84-87: Carl F. Roessler. 88: Shin Yoshino/Orion Press. 89: Mitsuaki Iwago/Orion Press. 90-91: David R. Gray. 91 Top: Wolfgang Bayer; bottom: Stewart Cassidy. 92-93: Tom Myers. 94: Teiji Saga. 94-95: Mark J. Rauzon. 96-97: Leonard Lee Rue, III. 98 Top: James A. Sullivan; bottom: Leonard Lee Rue, III. 99: Glenn D. Prestwich.

The Quest for Food
Pages 100-101: Tom Bledsoe. 103: Karl Weidmann. 104-105: Norman Myers/ Bruce Coleman, Inc. 106-107: George H. Harrison. 108-109: Jeffrey O. Foott. 109: Dr. Frank B. Gill. 110 Top: George F. Stover; bottom: Zig Leszczynski/ Animals Animals. 111: Zig Leszczynski/ Animals Animals. 112 Top: Jeffrey O. Foott; bottom: Ruth Smiley. 113: James H. Carmichael, Jr. 114: Robert B. Smith. 115: Gary R. Zahm. 116-117: Michael S. Quinton. 118-119: Jean-Paul Ferrero/ Ardea Photographics. 119: Teiji Saga. 120-121: Hans Reinhard/Bruce Coleman, Inc. 121 Left: Olive Glasgow; right: Bruce Hayward. 122: Alvin E. Staffan. 123 Top: Robert P. Carr; bottom: Constance P. Warner.

The Need for Shelter
Pages 124-125: Des and Jen Bartlett. 127: Terry Ator. 128-129: Karl H. Maslowski. 130: George W. Hornal. 131: Udo Hirsch. 132-133: George K. Bryce/Animals Animals. 133 Left: Jeffrey O. Foott/Bruce Coleman, Inc.; right: Hans Pfletschinger/Peter Arnold, Inc. 134: Christina Moats/Earth Images. 135: Willis Peterson. 136-137: Caulion Singletary. 138: Terry L. Schocke. 139: Michael Morcombe.

The Mating Season
Pages 140-141: Michael Morcombe. 143: James Tallon. 144-145: Steven C. Kaufman. 145: Glen T. Cheney. 146-147: C. C. Lockwood/Photo Researchers, Inc. 148-149 Top: Steven C. Wilson; bottom: Jeffrey O. Foott. 150-151: Mrs. Lorrimer Armstrong. 152-153: Fred Bavendam/Peter Arnold, Inc. 154: Mark J. Rauzon. 154-155: Dr. M. Philip Kahl. 156: William C. Gause. 157: Tsuneo Hayashida/Orion Press.

Threat and Rescue
Pages 158-159: Howard Hall. 161-163:
Medford Taylor/Black Star. 164-165:
Rajesh Bedi. 166-167: John Gronert.
168: Argus-Africa/Photo Trends.
168-169: Tom Myers. 170: IN THE
SHADOW OF MAN, © 1971 by Hugo
and Jane van Lawick-Goodall. 171: Top:
George and Lory Herbison Frame;
bottom: Hugo van Lawick, courtesy
National Geographic Society. 172-173:
Warren and Genny Garst/Tom Stack &
Associates. 174-175: Joseph S. Rychet-
nik. 176 Left: Don Halloran; right: Lynn
Rogers and David Buetow. 177: Laura
Medved. 178: George H. Harrison.
179 Left: George H. Harrison; right:
Dwight R. Kuhn. 180-181: Michael
Tennesen. 182-183: Jacques-Yves
Cousteau. 184: Ardea London. 185:
Rod Fox.

The Circle Closes
Pages 186-187: Tom McHugh/Photo
Researchers. 189: Karl Maslowski. 190-
191: Norman Myers/Bruce Coleman,
Inc. 192: Gene Hornbeck. 193: James
Pickerell/Black Star. 194-195: Top:
Erwin A. Bauer; 195 Bottom left: Michel
Folco/Black Star; 195 right: Jerome J.
Montague. 196-197: John Shaw. 197:
Dr. E. R. Degginger. 198: Pete Laurie/
S.C. Wildlife & Marine Resources Dept.
199 Top: George H. Harrison; bottom:
Smithsonian Institution Photo. 200: Dr.
E. R. Degginger. 201: Alan Root. 202-
203: Joe McDonald.

Library of Congress
Cataloging in
Publication Data

Main entry under title:

A Wildlife family album.
   Includes index.
   1. Zoology—Pictorial works.
I. National Wildlife Federation.

QL46.W58   599'.0022'2   81-81904
ISBN 0-912186-41-0   AACR2

Acknowledgments

Anyone who has assembled a family
album knows what fun it is to pore over
the pictures. However, getting all the
facts straight—everyone's name spelled
right, who is related to whom, the when,
why, and where each picture was taken
—can be work and require collaboration.

In writing the captions for this album,
we found that wild animals have a host
of friends who are ready to share their
latest observations of that vast, ever-
changing family. Among the many
specialists who helped us, we want espe-
cially to acknowledge generous help
from the following persons:

John G. Casey, National Marine Fisheries Service of the National Oceanic
and Atmospheric Administration, U. S.
Department of Commerce. Craig Phil-
lips, National Aquarium; Dr. Howard W.
Campbell, Denver Wildlife Research
Center; and Dr. Paul Opler, Office of
Endangered Species, all in the U. S. Fish
and Wildlife Service, U. S. Department
of the Interior. Jack Traub, California
Department of Fish and Game.

Dr. Wendall Swank, Texas A. & M.
University; Dr. Charles J. Jonkel, Uni-
versity of Montana; and Dr. Stephen
W. Kress, Cornell University.

Dr. Charles W. Myers, The American
Museum of Natural History. Dr. F.
Wayne King, Florida State Museum.
Ronald I. Crombie, Department of Her-
petology; Michael Carpenter, Depart-
ment of Invertebrate Zoology; William
A. Xanten and Charles Pickett, National
Zoological Park, all of the Smithsonian
Institution.

John C. Walsh, International Society
for the Protection of Animals; and Dr.
John Beecham, Bear Biology Association.

Some of the most fascinating infor-
mation came from the photographers
who relayed their own perceptions of
what the animals were doing before and
after as well as while their pictures were
being taken. We thank them all.

As always, we received support from
our NWF colleagues, naturalist Craig
Tufts, Raptor Center director William
Clark, and the editorial staffs of *Ranger
Rick*, *National Wildlife*, and *Interna-
tional Wildlife* magazines.

National Wildlife Federation

1412 16th St., N. W.
Washington, D. C. 20036

Jay D. Hair
Executive Vice President

James D. Davis
*Senior Vice President
Membership Development
and Publications*

Staff for this Book

Alma Deane MacConomy
Editor

Barbara Peters
Associate Editor

David M. Seager
Art Director

Dr. Raymond E. Johnson
Wildlife Consultant

Jo Ann Giliotti Smith
Research Editor

Rosa K. Hudson
Editorial Assistant

Priscilla Sharpless
Production and Printing

Kimberly Kerin
Catherine Yeardley
Production Artists

Margaret E. Wolf
Permissions Editor